Float or Sink?

Talya Jones

REAL LIFE readers™

Rosen Classroom™

New York

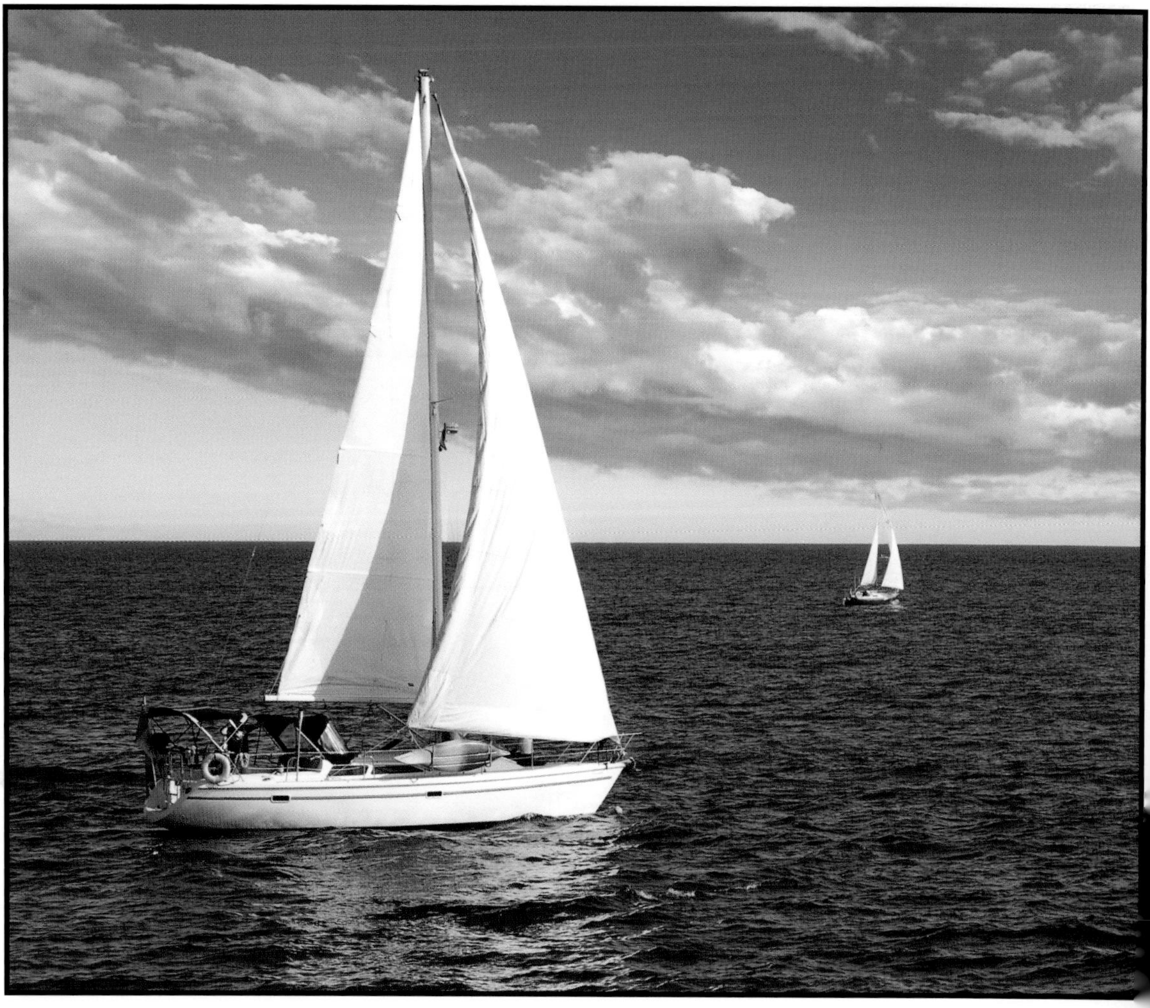

Boats float.

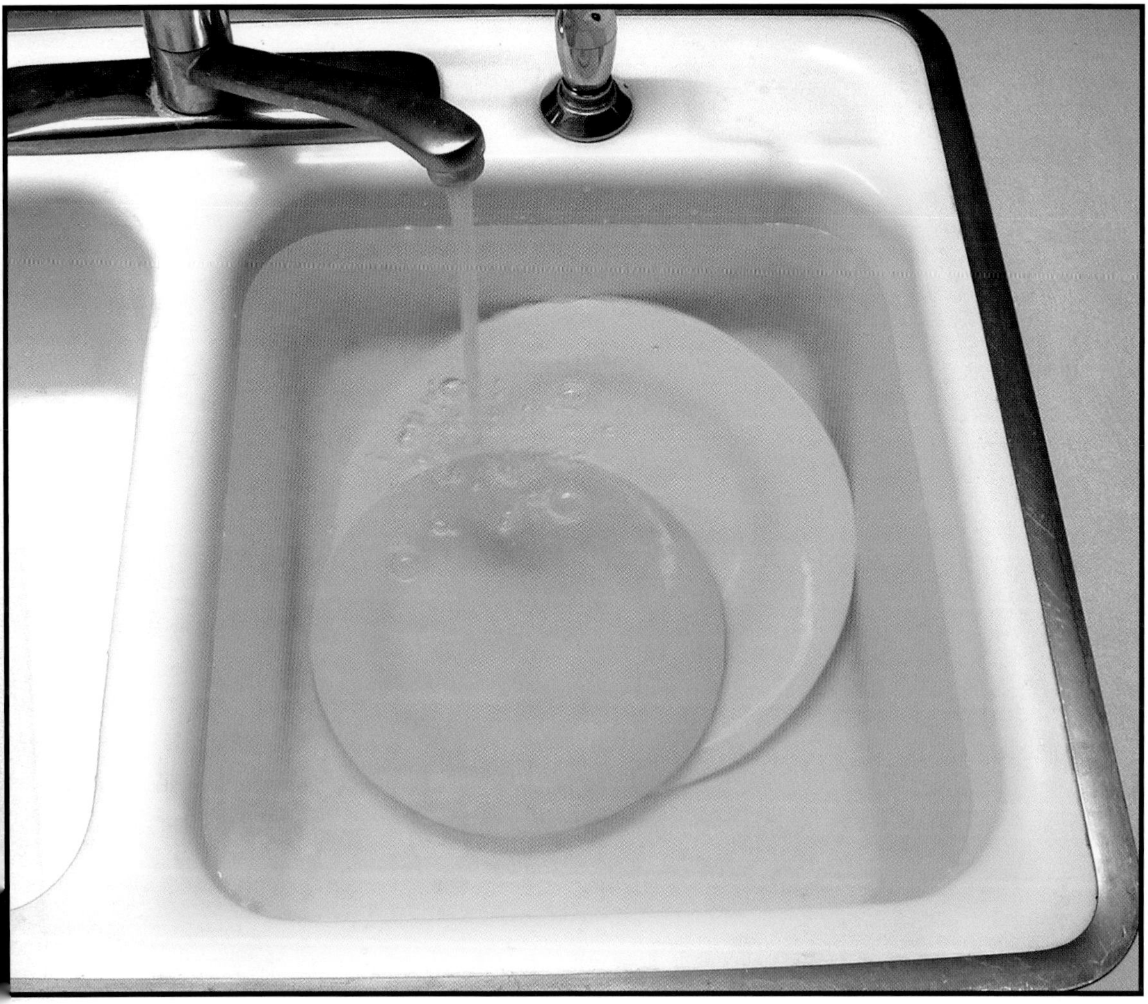

Dishes sink.

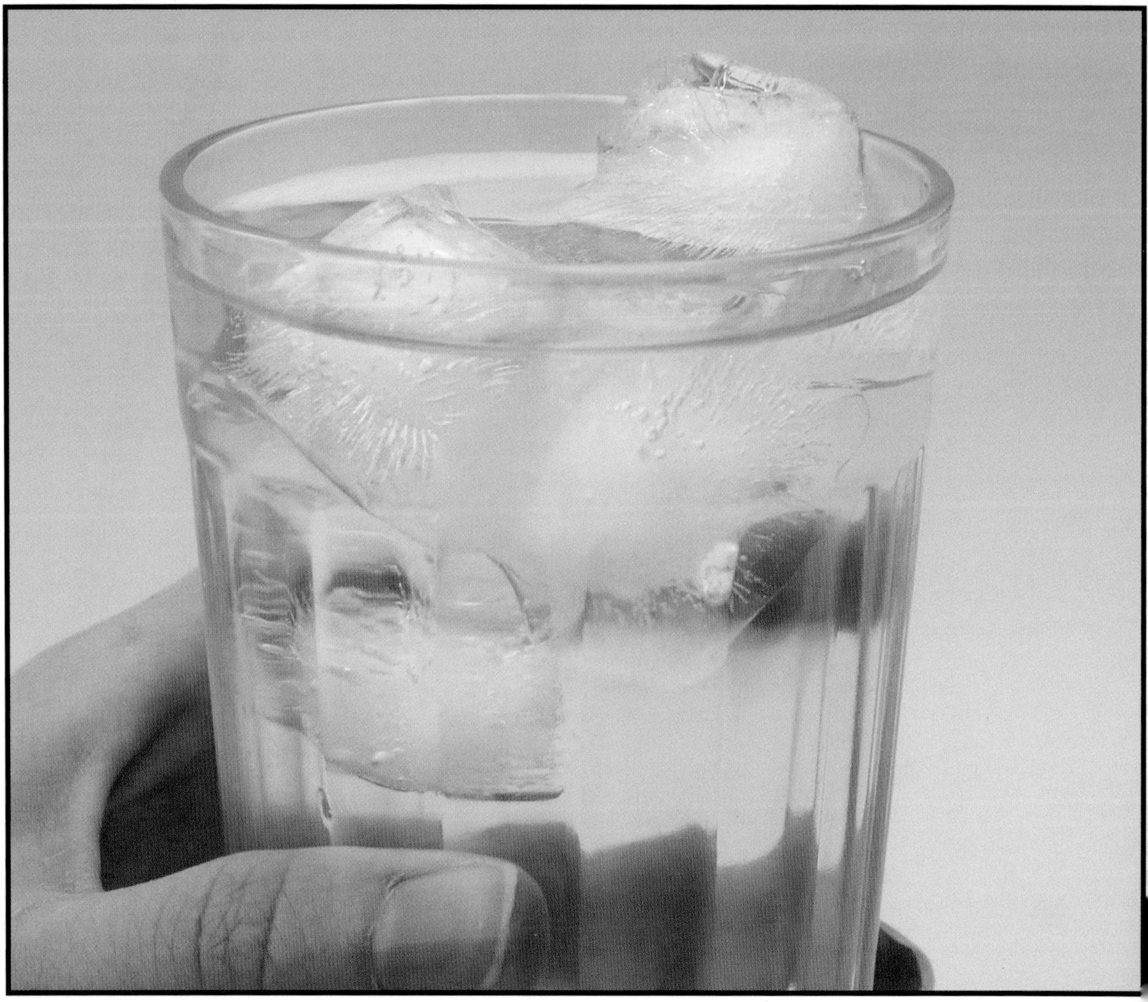

Ice floats.

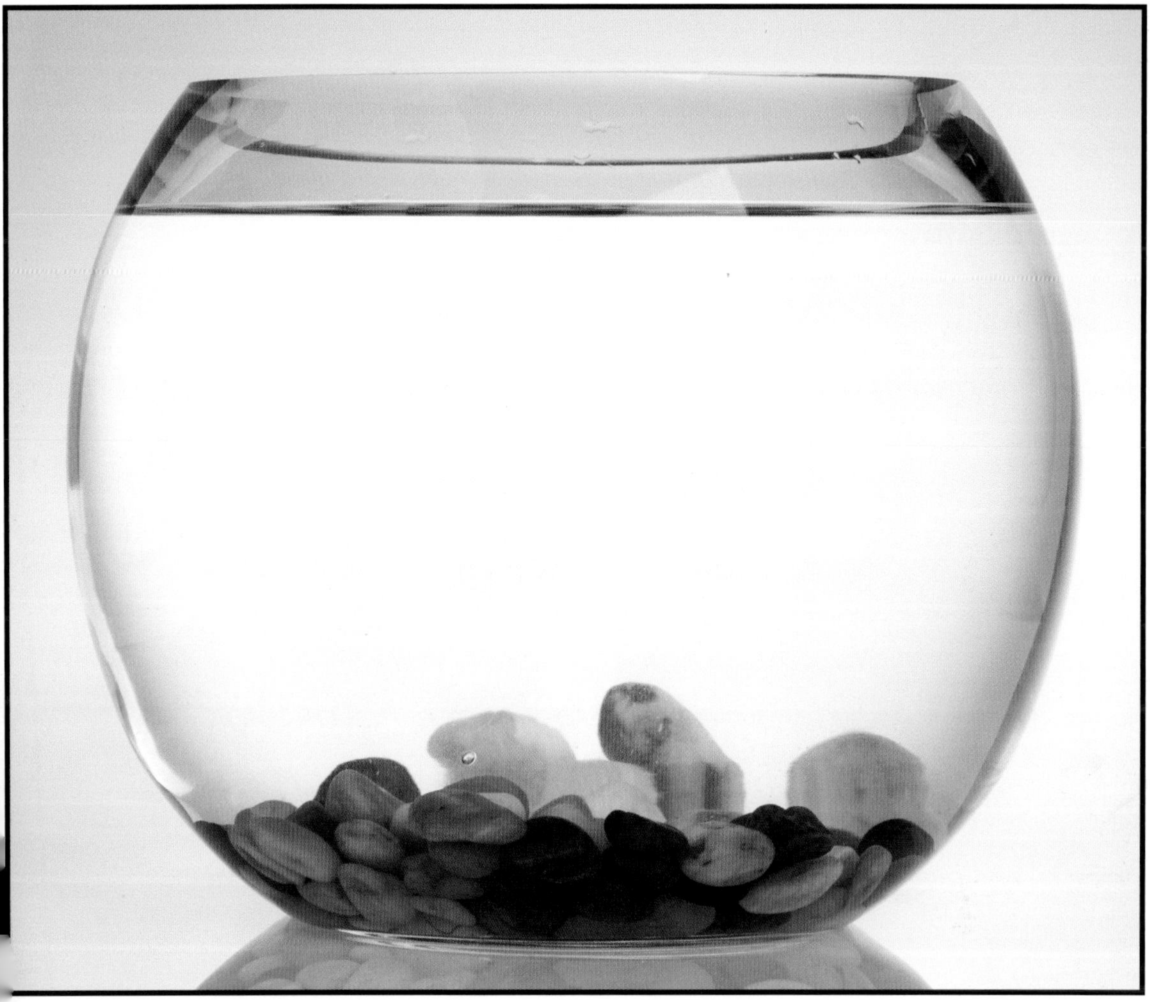

Rocks sink.

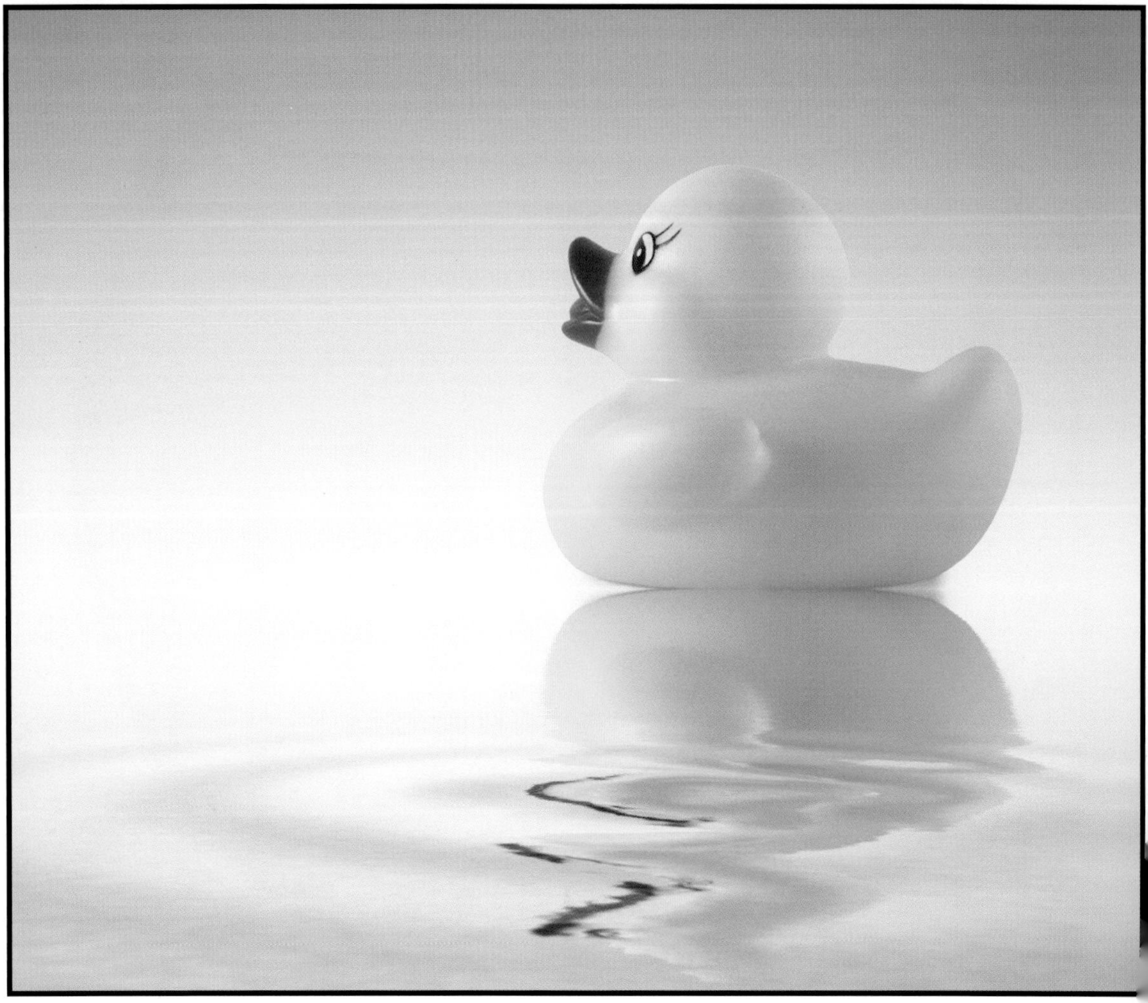

The toy floats!

Float	Sink
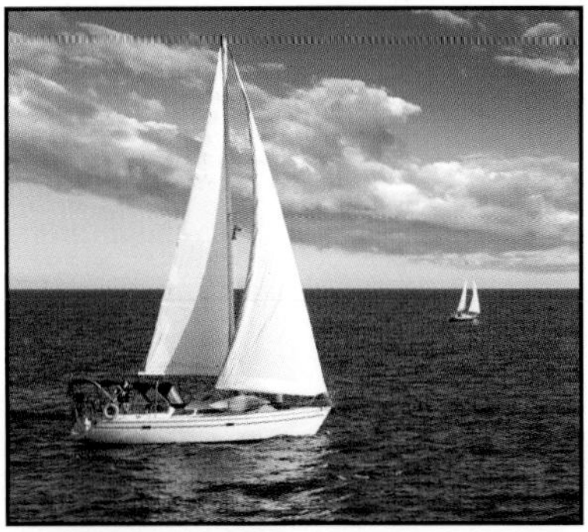	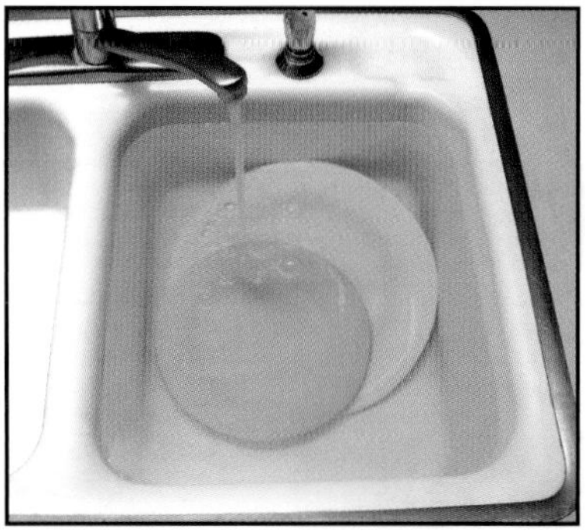
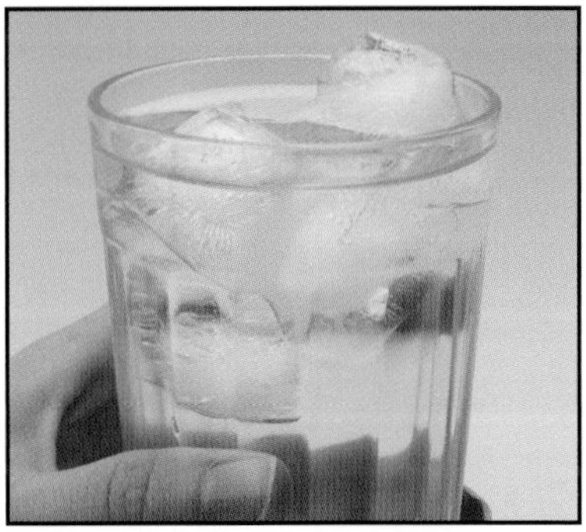	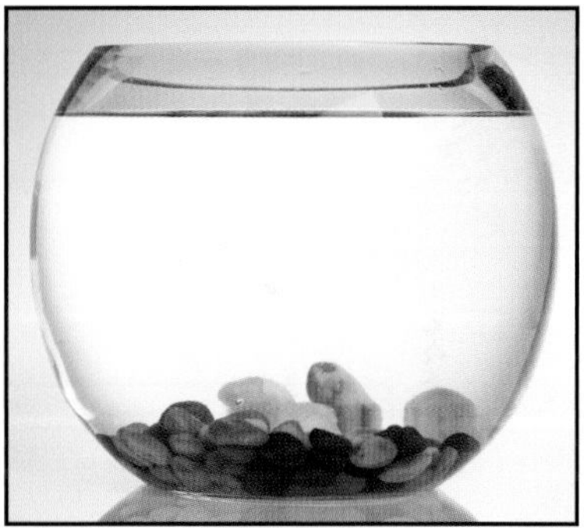
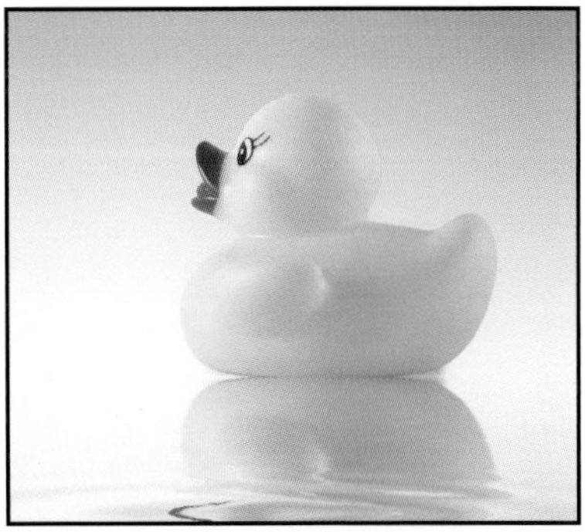	

Words to Know

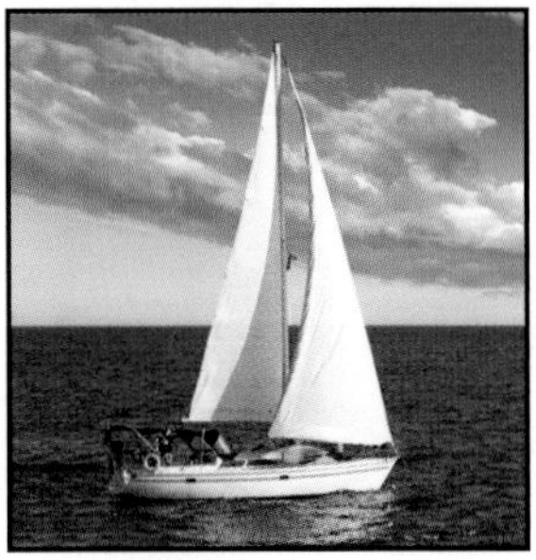

boat

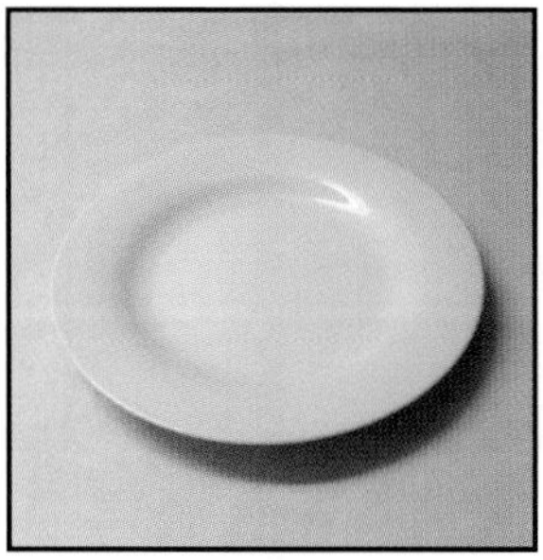

dish

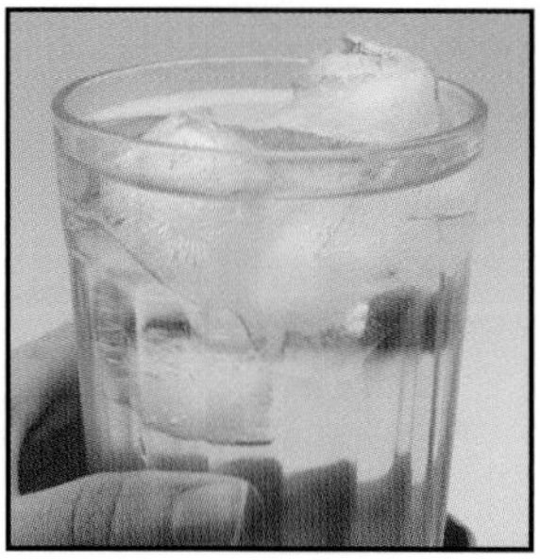

ice

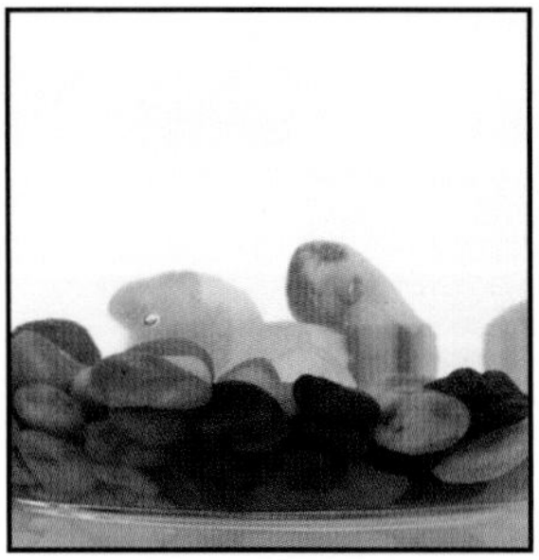

rocks

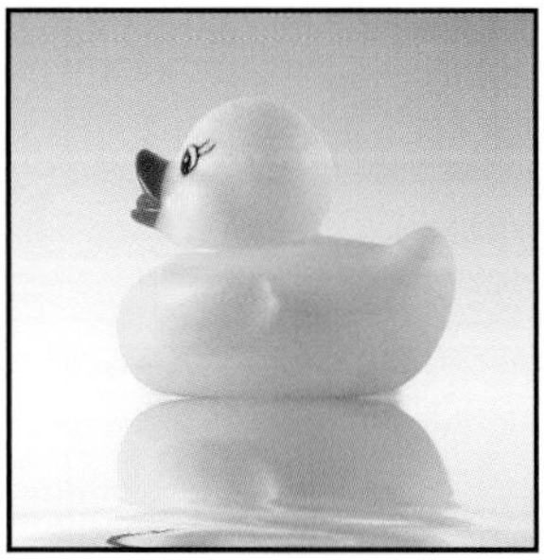

toy